# Mass Extinctions: Nature's Spectacular Staging of Natural Selection

Julian Lieb MD

ISBN: 1461184290

ISBN 13: 9781461184294

Library of Congress Control Number: 2011908181
CreateSpace, North Charleston, SC

## PREFACE

"A physician's subject of study is necessarily the patient, and his first field for observation is the hospital. But if clinical observation teaches him to know the form and course of diseases, it cannot suffice to make him understand their nature; to this end he must penetrate into the body to find which of the internal parts are injured in their functions. That is why dissection of cadavers and microscopic study of diseases were soon added to clinical observation. But today these various methods no longer suffice; we must push investigation further and, in analyzing the elementary phenomena of organic bodies, must compare normal with abnormal states. We showed elsewhere how incapable is anatomy alone to take account of vital phenenoma and we saw that we must add study of all physico-chemical conditions which contribute necessary elements to normal or pathological manifestations of life. This simple suggestion already makes us feel that the laboratory of a physiologist-physician must be the most complicated of all laboratories, because he has to experiment with phenomena of life which are the most complex of all natural phenomena."

*Claude Bernard*

*An Introduction to the Study of Experimental Medicine (1865), trans.*

*Henry Copley Green (1957), 43*

In Lily Splane, Quantum Consciousness (2004), 307

Science quotes on: <u>Science And Art</u> (32)

Please consult a biomedical database for references.

# STRANGE MASS DEATHS

In 2006, a mass death of honeybees and bats was observed; the latter referred to as the white nose syndrome and attributed to a fungus, and followed by mass deaths of pilot whales, pelicans, and sea turtles. In December, 2010, thousands of dead fish washed up in Australia, in Florida, and the Philippines, more than a hundred pelicans died in North Carolina, and ten tons of dead fish washed up in New Zealand.

Thousands of dead red winged blackbirds fell from the Arkansas sky on New Year's Eve 2010, followed by 2 million dead fish washing up in Chesapeake Bay, 150 tons of dead red tilapia in Vietnam, and 40,000 dead crabs in Britain. Then 900 turkey vultures "drowned and starved," 4,300 ducks were "killed by parasites," 1,500 salamanders were terminated by a "virus" and 2,000 bats died from "rabies".

In the months that followed, mass deaths occurred in pelicans, fish, manatees, bumble bees, buffalo, water buffalo, cows, flamingoes, frogs, sardines, anchovies, manta rays, dolphins,

1

whales, baby right whales, ducks, clams, octopuses, lobsters, peacocks, seals, budgerigars, earthworms, crickets, prairie dogs, sharks, forests, sea grass, squid, and grassland butterflies. Oysters, French standard poodles, elk, deer, moose, and wild boars perished, while in the aftermath of the Joplin tornado, a few injured people developed a rare and potentially fatal fungal infection, so aggressive it turned their tissues black and grew inside their wounds, a similar cluster never seen before. Episodes of mass deaths continued. In June, 2011, 800 tons, or 1.6 million pounds of fish washed up dead in a lake south of Manila. We will never know the full extent of the mass deaths, and perhaps it's for the better.

## STRANGER INTERPRETATIONS

Officials attributed the mass deaths to oxygen levels in the world's oceans dropping dramatically, extreme cold weather, Arctic outbreaks and heavy snows, farmers allowing fish pens to become overcrowded, ebb tides causing the oxygen level in the water to drop, commercial fisherman dumping their catch, and broken fish nets. Penguins washed ashore, along with other shorebirds, interpreted as fair-weather high-pressure parking overhead, reducing the small aquatic animals that fish feed on, forcing fish to move to find food, thus leaving the penguins and shorebirds to starve to death. Thousands of turtle doves dropped dead from "massive indigestion" the birds "gorging themselves" on sunflower seeds from a nearby sunflower oil factory, the overeating damaging livers and kidneys. Cows died from eating "spoiled sweet potatoes" birds from "eating snowmelt," or dropping dead en masse "scared to death" by "sound of firecrackers."

A fungus is said to be at fault for the deaths of 1 million North American bats from the "white nose syndrome," according to a study in Nature magazine. Bats eat about two-thirds of their own weight in insects nightly that can spread disease and ruin crops. The researchers noted healthy bats exposed to the fungus "got the syndrome," and that sick bats could infect healthy ones through "physical contact" during hibernation, when they congregate in large numbers in caves. All at the same moment in time, infecting each other synchronously, by rubbing their noises together? The white nosed syndrome the only mass death not induced by natural selection?

Oddly enough, Charles Darwin and Alfred Wallace uncovered the truth for the strange events, Darwin publishing it one hundred and fifty one years ago in his "Origins of Species," and referred to it as "natural selection."

## NATURAL SELECTION

One hundred and fifty one years ago, Charles Darwin and Alfred Russell Wallace would have known that birds dropping from the skies; dead fish, crabs, and sardines washing up from the sea; dead cows and water buffalos dropping in the fields; and dead bats dropping in caves and from under bridges, are nature's spectacular staging of what Darwin termed natural selection. Darwin wrote, "As many more individuals of each species are born than can possibly survive; and as, consequently, there is a frequently recurring struggle for existence, it follows that any being, if it vary however slightly in any manner profitable to itself, under the complex and sometimes varying conditions of life, will have a better chance of surviving, and thus be naturally selected. From the strong principle of inheritance, any selected variety will tend to propagate its new and modified form."

"It may metaphorically be said," Darwin wrote, *"that natural selection is daily and hourly scrutinizing, throughout the world, the slightest variations; rejecting those that are bad, preserving and adding up all that are good; silently and insensibly working, whenever and wherever opportunity offers....We see nothing of these slow changes in progress, until the hand of time has marked the lapse of ages, and then so imperfect is our view into long-past geological ages, that we see only that the forms of life are now different from what they formerly were."*

Wallace wrote, "An antelope with shorter or weaker legs must necessarily suffer more from the attacks of the feline carnivora; the passenger pigeon with less powerful wings would sooner or later be affected in its powers of procuring a regular supply of food . . . If, on the other hand, any species should produce a variety having slightly increased powers of preserving existence, that variety must inevitably in time acquire a superiority in numbers. . . . Now, let some alteration of physical conditions occur in the district — a long period of drought, a destruction of vegetation by locusts, the eruption of some new carnivorous animal seeking "pastures new" . . . it is evident that, of all the individuals composing the species, those forming the least numerous and most feebly organized variety would suffer first, and, were the pressure severe, must soon become extinct."

# PROSTAGLANDINS: MOLECULES OF NATURAL SELECTION AND EVOLUTION

## THE ARACHIDONIC ACID CASCADE

Bengt Samuelsson (Nobel Prize winner, 1982)

An elucidation of the arachidonic acid cascade.

Discovery of prostaglandins, thromboxane and leukotrienes.

Drugs. 1987; 33 Suppl 1:2-9.

## Abstract

Arachidonic acid is normally stored in membrane-bound phospholipids and released by the action of phospholipases. Enzymatic conversion of released arachidonic acid into biologically active derivatives proceeds through one of several routes. Cyclooxygenase converts arachidonic acid to unstable cyclic endoperoxides from which prostaglandins, prostacyclin and thromboxanes are derived. Formation of leukotrienes from arachidonic acid is initiated by the action of 5-lipoxygenase. Prostaglandin E2 and prostacyclin are potent vasodilators, while leukotriene D4 causes cellular adhesion and movement of white blood cells. Leukotrienes C4, D4 and E4 contribute to inflammation by increasing vascular permeability, leukotrienes playing an important role in allergic asthma. Through pharmacological intervention in the arachidonic acid cascade various anti-inflammatory agents have been developed. These include aspirin-like drugs, which inhibit cyclooxygenase. Corticosteroids appear to indirectly inhibit phospholipases thus preventing release of arachidonic acid. Future progress in this field is likely to produce drugs which antagonize arachidonic acid derivatives or inhibit the enzymes involved in their synthesis with greater specificity."

Bengt Samuelsson

The products of arachidonic acids are referred to as eicosanoids. They regulate cells producing them and adjacent cells, are rapidly degraded and not carried in the bloodstream to distal organs as hormones. Examples of eicosanoids are prostaglandins, prostacyclin, thromboxanes, and leukotrienes. Prostaglandins are ephemeral, infinitesimal, and powerful molecules that signal throughout each cell, and from cell to cell, organ to organ, brain to body, body to brain, and environment to body. They weigh one-trillionth ($10^{-12}$) of a gram. Their transient existence, speed, production and destruction are paradoxically responsible paradoxically for cellular function and dysfunction. roles in inflammation, fever, regulation of blood pressure, blood clotting, and immune system modulation, control of reproduction and tissue growth, and regulation of the sleep/wake cycle. Leukotrienes are involved in asthma and psoriasis, and in the calcified heart valves of aortic stenosis, known to be similar to atherosclerosis (hardening of the arteries). Draw blood and measure its prostaglandins; now gently tilt the tube, measure again, and you will find a marked increase in the level. This is why it is presently impossible to use blood levels for clinical purposes.

In 1980, Dr A. Bennett and his team examined tumor-associated prostaglandin-like material in 37 patients with primary and metastatic carcinomas of the head and neck, previously treated by radiotherapy and chemotherapy followed by radical surgery. High amounts of prostaglandin-like material were extracted from tumors excised within 3 months of radiotherapy and chemotherapy. Some would argue that the process of extraction might impact prostaglandin levels, but this was a consistent finding across 37 patients. Tradition, common usage and the medical literature favors "prostaglandins" as a generalization over "eicosanoids."

In the early 1930's, Maurice Goldblatt in the United States and Ulf von Euler in Sweden showed that factors in the seminal fluid of boars act on various smooth muscles and lower blood pressure. Von Euler named these substances "prostaglandins" because the prostate contains small amounts of them, and he assumed that what he extracted from semen must have come from that gland. Today we know that every cell manufactures prostaglandins, or other members of the eicosanoid family.

At the Karolinska Institute in Stockholm, Sune Bergstrom purified several prostaglandins, determined their chemical structure, and showed that they are formed from essential fatty acids. After collaborating with Bergstrom from 1959-1962 on the structure of prostaglandins, Bengt Samuelsson provided a detailed picture of arachidonic acid and prostaglandin metabolism, and defined the chemical processes involved in their synthesis and breakdown. Samuelsson showed that blood platelets convert arachidonic acid into thromboxanes, while white blood cells convert it to leukotrienes. Thromboxanes constrict blood vessels and cause platelets to clump together and release more clotting factors. This is useful when clotting is necessary to stop bleeding; when this mechanism is overactive, it plays a pivotal role in heart attacks and strokes. Thromboxanes directly stimulate the smooth muscles of blood vessels to contract, including those of the heart and brain.

Oxford University's Sir John Vane showed that anti-inflammatory compounds such as aspirin block the formation of prostaglandins and thromboxanes. Later Vane and Salvador Moncada isolated a prostaglandin in the wall of blood vessels and named it prostacyclin. In dilating blood vessels and inhibiting the aggregation of platelets, prostacyclin opposes the actions of thromboxanes. When prostacyclin was injected into mice, it induced a syndrome resembling major depression in humans.

Prostaglandins are known to have played essential roles in the evolution of coral, sponges, algae, sea cucumbers, shellfish, fish, frogs, fungi, insects, reptiles, vertebrates, mammals, primates and humans, but never before in natural selection. Prostaglandins regulate the synthesis, activation, release and degradation of enzymes, and the synthesis, inhibition, and expression of deoxyribonucleic acid (DNA).

The preimplantation human embryo would be destroyed by an attack by the mother's immune system, as half of its contents come from its father. Instead, the embryo temporarily increases its production of prostaglandin E2, which, when elevated, becomes a powerful immunosuppressant, and wards off the attack. In the absence of this mechanism, reproduction would be impossible. If the enzymes that produce prostaglandins increase their production slowly, aging will be slow, and those so fortunate unlikely to develop prostaglandin - generated disorders that shorten our lifespan, centenarians virtually immune to medical and surgical disorders.

Prostaglandins determine tolerance or intolerance towards everything with which the body comes into contact, among them temperature fluctuations, venom, emotional stress, shear stress, microgravity, oxidative stress, changes in ionic composition, medications, alcohol, allergens, carcinogens, euphoriants, gases, humidity, light, dark, sound, electromagnetic fields, water, microorganisms and food. Electromagnetic fields regulate enzymes, directly and indirectly, by acting on cell membranes. Prostaglandins orchestrate cognitive, emotional, behavioral, physiological, pathological, and reproductive responses to the environment.

Prostaglandins are not produced under resting conditions, but only in response to stimuli. If the enzymes that produce and degrade prostaglandins are resilient, they can absorb stress, and continue to function physiologically; if not, physiology becomes pathology. People with well- regulated prostaglandin production are relatively immune to stress, people lacking such control hypersensitive, with those with depression at the top of the list. Virtually every medicine ever invented interacts with prostaglandins. Aspirin and ibuprofen are appreciated as inhibiting prostaglandins, lithium and antidepressants generally unknown in this context.

Nothing excites the curiosity of pharmacologists more than the discovery of a previously unknown family of molecules, and this held true for prostaglandins. In the early nineteen seventies, researchers showed that antidepressants inhibit the mobilization of arachidonic acid, have agonist/ antagonist actions on prostaglandins, and reduce the brain levels of depressant prostaglandins. Over the past decade, scientists at the National Institutes of Health have shown that lithium reduces brain phospholipases and cyclooxygenase activity, as well as prostaglandin E2 concentrations in rat brains.

15- HD is the primary prostaglandin-degrading enzyme, highly expressed in normal colon mucosa but lost in human colon cancers. Lack of this enzyme promotes the earliest steps of growth of benign as well as malignant colon tumors. A reduction in this enzyme may elicit an increase in cyclooxygenase, which is not good news. When this enzyme was first characterized, every agent tested in the hope of stimulating it either had no effect or inhibited it. Eventually Mak and Chen showed that tricyclic antidepressants powerfully activate the enzyme in mice, especially the kidney enzyme, one of them with more than a thousand -fold activation, and with potent activating effects on this enzyme in the brain.

Increased synthesis by enzymes prostaglandin E2, or inhibition of its degradation beyond a critical threshold, is responsible for depression on the one hand, and for defective immunity, autoimmunity, cancer and neurodegenerative disorders on the other. Excessive thromboxane B2 is responsible for such cardiovascular disorders as heart attacks and strokes. Antidepressants are often effective in preventing, alleviating, or reversing these disorders. They inhibit the enzymes that synthesize prostaglandins, and stimulate those that degrade them. These mechanisms illuminate the causes of these disorders, but not their variations. Why does one person suffer from depression and heart disease, another from depression and recurrent infections, a third from depression and various autoimmune disorders? The answer lies at the interface between carboxylic acids (prostaglandins) and nucleic acids (genes). Why would excessive production of prostaglandins in the brain account for defective immunity and autoimmunity? Enzymes function paradoxically. Why do antidepressants both alleviate disorders of defective immunity and autoimmunity? Antidepressants have paradoxical actions on prostaglandins. Lithium inhibits the turnover of arachidonic acid by down regulating brain phospholipases, and thus has potent immunostimulating, antiviral, and anti-bacterial actions.

Databases contain copious evidence of the role of prostaglandins in virtually all diseases and disorders. This constellation can be explained only by regulation of the arachidonic acid pathway. When the human genome was found to contain fewer genes than could account for human diversity, geneticists conjured up "epigenetics" and then "dancing genes" to preserve the dominance of genes, while ignoring the earlier studies of Horrobin and Hughes-Fulford.. Excessive synthesis of prostaglandins converts physiology to pathology, the variations of the disease or disorder determined by genes. Georgia tech biologist John McDonald proposes that the structural and behavioral differences between humans and chimpanzees are mainly due to differences in the regulation of genes rather than to differences in their sequence.

In the nineteen-seventies-and-eighties, prostaglandins attracted substantial drug company investment, one of which referred to them as "Medicine's New Frontier." With advent of technology that accelerates the production of DNA, venture capitalists, the U.S patent office, medical schools, and media launched biotechnology, stampeding these companies into divesting from prostaglandins, and purchasing biotechnology companies, to acquire their patents. In 1980, the first genomics company was launched under the banner of a virtual cure- all for disease, plausible to the uninformed, but leaving the informed devastated knowing that genes could not be accessed for clinical purposes. Proteinomics and stem cells followed, with prostaglandins missing in action. Many individuals and institutions purged prostaglandins so thoroughly, that they almost disappeared from sight and mind.

Medical research is largely based on the premise that DNA and RNA in the cell nucleus, and enzymes and proteins in the cell, whose structures are defined by DNA, are of overwhelming importance in disease. Membrane lipids, which regulate these entities, offer far more opportunities for practical therapeutic interventions. In 1893, Sir Norman Locklear postulated that the cell membrane has a major role in cellular metabolism, the nucleus a distinctly minor one,

## AN EXPERIMENT ON INFERTILITY ILLUMINATES PROSTAGLANDINS IN NATURAL SELECTION

In the mid-1920s obstetrician Raphael Kurzrok noted that "out of dozens of attempts at artificial insemination, only two were probably successful. In a number of cases he observed that when 0.5 ml of semen was injected into the uterine cavity, the semen was promptly expelled. A similar quantity of Ringer's solution similarly injected was invariably retained. The patient always had the same reaction, apparently independent of the phase of the menstrual cycle". Kurzrok and colleague Charles Lieb suspended strips of uterine muscle in 100 ml of warm, oxygenated Ringer's solution, to which they added 1 ml of warm semen. They write: "The same uterus may react to semen by contracting; to another by relaxation. The same semen may contract one uterus and relax another. From this we may draw the tentative conclusions that certain types of sterility are sometimes due to the female, sometimes to the male. A study of the history of the patients from whom the uterine strips were obtained throws an interesting light on our experiments. The uteri from the patients who give a history of successful pregnancy responded to fresh semen by relaxation, while uteri from women who gave a history of complete or long-standing sterility were always stimulated by semen".

The molecules in semen responsible for these reactions are now known to be prostaglandins. Reproduction and survival are the cornerstones of natural selection, and prostaglandins are ubiquitous in both. As immune regulators, prostaglandins have an essential role in survival by preserving health or, paradoxically, inducing defective immunity, autoimmunity, and cancer. Kurzrok and Lieb's experiment illuminates prostaglandins as molecules of natural selection.

# CARBOXYLIC ACIDS (PROSTAGLANDINS) AS NATURAL SELECTION, NUCLEIC ACIDS (GENES) AS VARIATION

"The absence of a coherent alternative to Neo-Darwinism makes many biologists feel that a bad theory is better than no theory at all."

Arthur Koestler,

"Does a living thing somehow have within itself, consciously or unconsciously, a way of responding actively to the environment and shaping the future of its species accordingly? If so, it would go a long way towards solving the two basic conundrums of evolution...its speed, and the need for a number of mutations to take place simultaneously. Can we find, within an organism, a response system that accelerates genetic change in this way?"

Francis Hitching

...No one has so far synthesized anything as remotely complex as an enzyme or any other protein molecule.... These complex molecules do not simply assemble themselves from mixtures of ingredients like a cup of tea. Something else is needed. What the something else is remains conjectural. If it is a chemical it has not been discovered; if it is a process it is an unknown process; if it is a 'vital principle' it has not yet been recognized. Whatever the something is, it is presently impossible to build a case for Darwinism or against vitalism out of what we have learned of the cell and the molecules of which it is composed."

Richard Milton

Neo-Darwinism emerged as an attempt to explain natural selection by combining algebra and Mendelian genetics. If small variations are the result of changes in genes (DNA) they are preserved. If dominant, they will recur in subsequent generations; if recessive, they will remain dormant until two people with the same gene mate. In ultra Darwinism the gene is the only level at which selection occurs, natural selection is the only force driving evolution, organisms are infinitely flexible to change, and the purpose of life is the reproduction of genes.

In an article entitled "Some biologists ask, are genes everything" in the September 2, 1997 edition of the New York Times, Sandra Blakeslee interviewed University of Sussex biologist Brian Goodwin. Goodwin commented, "A gene makes a protein and that's about it. It doesn't tell you how proteins interact, how cells and tissues communicate, how organs come into being, how an immune system forms, or how evolution works... We need to discover what brings about changes in form."

I knew that prostaglandins regulate cell-to-cell signaling, the shape and form of cells, and other mechanisms involved in the formation of the embryo and the immune system. Immersing myself in the evolutionary literature and drawing on clinical observation, I learned that prostaglandins fill the gaps defined by Goodwin. When I read that natural selection depends on replication and reproduction, I knew I was on track. Contrary to received wisdom, prostaglandins and not nucleic acids are responsible for replication. As for reproduction, the challenge is to identify mechanisms not regulated by prostaglandins, rather than those that are.

When I studied the natural selection literature, I realized its lack of a credible biochemistry. At first I suspected that prostaglandins are the replicators and interactors of natural selection. Then I realized that natural selection is a property of prostaglandins, as are such mechanisms of evolution as variation, mutation, homology, neoteny, heterochrony, aging, apoptosis, infection, autoimmunity, cancer, aging, death and extinction. Finally, I concluded that prostaglandins are natural selection, and genes variation.

Marc Lappe, Paul Ewald, Ralph Nesse and George Williams pioneered the concept of evolutionary medicine. I had the advantage of studying prostaglandins, and knowing that they differentiate between reproduction and sterility, health and death. As a clinical researcher, I observed the response of infections to lithium and antidepressants, of cancer and autoimmune disorders to antidepressants and knew that these agents inhibit prostaglandins.

# THE MOLECULES OF NATURAL SELECTION

## Molecules of Reproduction

The molecules of evolution are an integral component of biological clocks and preside over the machinery of reproduction. They participate in sexual maturation, ovulation, the menstrual cycle, erection, ejaculation, fertilization, implantation, embryogenesis, giving birth, neoteny (the existence of juvenile features in an adult) and heterochrony (changes in the timing or relative rates of development of different tissues within an organism that can be inherited by its offspring).

Prostaglandins regulate the machinery of reproduction, from erection and ejaculation to the delivery of the placenta. When infertile women were given 100 mg aspirin a day in addition to a fertility drug, their production of ova and rates of pregnancy increased by 50% over women given the fertility drug alone. Prostaglandins similarly govern reproductive success in males. If excessive prostaglandin production were common to infertility, depression and natural selection, one would anticipate an increase in infertility among depressives. In a retrospective study, Dr Helen Ossofsky identified irregular menses, lack of menses, premenstrual syndrome, infertility, spontaneous abortion and precipitate labor as part of the clinical picture of women with depression.

Moore and Waring have shown that the urine from ovulated mature female salmon contains a priming pheromone that enhances the reproductive physiology of receiving males. Electrophysiological studies have shown that the olfactory epithelium of mature male salmon is acutely sensitive to prostaglandin F1alpha and prostaglandin F2 alpha. Sensitivity to these prostaglandins increases as the reproductive season progresses.

The New Zealand reptile tuatara dig a nest over a period of several nights, lay a complete clutch of eggs on a single night, and then guard the nest for several nights. L J Guilette at the University of Florida has shown that prostaglandin E2 is elevated during nest digging and laying eggs, but declines during nest guarding. Plasma prostaglandin F is elevated during nest digging, rises significantly during laying of eggs, and declines to baseline levels during nest guarding. These data suggest that the role of prostaglandins in egg laying function may have been conserved throughout evolution of vertebrates that develop from an embryo within an amnion, e.g. a bird, reptile, or mammal.

## Molecules of Inheritance

The molecules of evolution regulate the synthesis and expression of genes, and the assembly of polypeptide chains into proteins and enzymes. They connect genes to the environment, and control the inheritance of mitochondria. Prostaglandins have these properties.

## Prostaglandins in embryogenesis (gastrulation)

In "The Developing Human: Clinically Oriented Embryology" Keith L. Moore and T.V.N. Persaud note, " The molecules of embryogenesis cause slight differences in the embryo in successive generations, altering not only the shape of an animal's body but also its behavior. These molecules regulate cell division and growth, the interaction of proteins, mass cell movements allowing cells to interact with each other, differentiation of cells into tissues and organs, and the development of form, formation and release of enzymes, regulation of size and programmed cell death."In 1981 Kenneth L Klein showed that prostaglandins are highly active in the embryo, and have an essential role in embryogenesis. Yi Cha has shown that prostaglandin E2 production is essential for embryogenesis movements in zebrafish.

## Molecules of Response to the Environment

The molecules of evolution regulate the response of cells to the external and internal environments. In 1978, P.D. Buisseret showed that doses of aspirin, indomethacin, or ibuprofen prevented symptoms of food intolerance in five out of six patients who on several occasions had had acute gastrointestinal symptoms after ingesting specific foods. Blood and stool prostaglandin E2 and F2α concentrations during unprotected challenge were consistent with the idea that these symptoms were mediated through prostaglandin release. This study should have formed the basis of a paradigm shift for food intolerance and allergy, but fell by the wayside while prostaglandins were suppressed in favor of biotechnology and genomics.

In Vietnam, Nguyen Thi Phuong went from a youthful beauty to one with the appearance of a seventy-year old, with sagging and wrinkled skin all over her face and body, following an allergic reaction to seafood. She noted that rapid aging had not affected her menstrual cycle, hair, teeth, eyes and mind. One suspects a prodigious increase in prostaglandin E2 instigated by the cyclooxygenase enzyme in some cell lines but not others as determined by nucleic acids.

Snake venom and the poison of sea nettles contain high concentrations of leukotrienes, while the saliva of ticks, fleas and mosquitoes is saturated with prostaglandins. Prostaglandins regulate the interaction between living beings and their environment. They are sensitive to movement, changes in ambient temperature, barometric pressure, gravity, light, food sources, gases, radiation, electricity and stress. The biological environment is itself largely controlled by prostaglandins. The outcome of any contact between microorganism and host is determined by the ability of the former to induce prostaglandin production, and of the latter to control it.

Ragweed allergy is mediated by leukotrienes, the venom of sea nettles, spitting cobras, sea snakes, honey bees, yellow jackets, and Russel vipers known to contain high concentrations of them. The saliva of ticks teems with prostaglandin s; when a tick inserts its mouthparts, prostaglandins depress local immune function and draw in blood supply, both conducive to infecting the host with such organisms as those causing Lyme disease.

Environmental agents (teratogens) may cause developmental problems if the mother is exposed to them. The outcome of exposure depends on critical periods of development, the dosage of the drug or chemical, and the constitution of the embryo. The heart, limbs and upper lip are most vulnerable during week's four to eight, the eyes and teeth from the ninth week until term, the central nervous system from week thirty two until term. Teratogenic drugs include androgens, high doses of progestogens, aminopterin, busulfan, cocaine, diethlstilbestrol, lithium carbonate, methotrexate, Dilantin, tetracycline, thalidomide, valproic acid, warfarin and minoprostil.

Among infectious agents that can cause congenital anomalies, prematurely or stillbirth when transmitted from mother to fetus are cytomegalovirus, herpes simplex virus, human immunodeficiency virus, human parvovirus B19, rubella virus (German measles), Toxoplasma gondii, treponema pallidum (syphilis) Venezuelan equine encephalitis virus, and the varicella virus ( chicken pox and shingles). Environmental teratogens include methlmercury, polychlorinated biphenyls, (PCB's), alcohol and ionizing radiation.

All of these agents induce the synthesis of prostaglandins. The ability of prostaglandins to cause birth defects first emerged with their incrimination in a heart defect known as patent ductus arteriosus. In the embryo the ductus arteriosus connects the left pulmonary artery with the aorta. During embryonic development prostaglandins maintain its patency. In normal infants the onset of respiration following birth increases the level of oxygen in the blood, reducing the level of prostaglandins in the ductus. In infants with patent ductus arteriosus, prostaglandin levels remain above this level and the ductus remains open. Surgical closure of the shunt was the standard procedure for the disorder until pediatricians discovered that small doses of such anti-prostaglandin drugs as indomethacin could close the ductus tangible evidence of an anti-prostaglandin drug reversing an evolutionary disorder.

## Prostaglandins in heterochrony

Heterochrony refers to a difference in the timing of formation of parts of tissues, or the occurrence of a phenomenon at an unusual time. The key mechanisms in heterochrony are acceleration and retardation. Organisms are mosaics in which each organ undergoes its acceleration or retardation at a different rate. Many biologists assume, as Darwin did, that natural selection acts mainly on late embryonic or postnatal development. As indicated below, prostaglandin heterochrony determines sex behavior.

The paradoxical actions of such prostaglandin - inhibiting agents as lithium and antidepressants offer examples of acceleration and retardation. Antidepressants can arrest rheumatoid arthritis, even reversing a deformed joint to its pristine condition. They can accelerate a slow cycling mood disorder into one that is rapid cycling, and arrest and accelerate viral replication. Antidepressants are capable of slowing aging, and inducing rejuvenation.

## Prostaglandins in neoteny

Neoteny refers to the ability of an immature phase to reproduce. When mussels and clams are cultivated in feeding beds some spawn when premature, others when mature and the balance when post-mature. When harvested only the mature are edible. When small doses of fluoxetine are added to the beds the bivalves spawn synchronously. Every antidepressant so tested inhibits prostaglandins.

## Prostaglandins in aging and recapitulation

Aging is an evolutionary process. Prostaglandins regulate aging, and along with it the prevalence of disorders of defective immunity and autoimmunity. As post mortem brains of patients with Alzheimer's disease contain abnormally high concentrations of prostaglandins, it is not surprising that ibuprofen reduces the risk of this disease. A slow increase of prostaglandins with age may explain why centenarians are relatively free of cancer and Alzheimer's disease. Conversely, progeria (premature aging) carries an elevated risk for autoimmune disorders.

Three women and two men treated for arthritis with the prostaglandin - synthesis inhibitor benoxaprofen, developed excessive and accelerated nail and hair growth. In both men the drug reversed hereditary male-pattern baldness. Curiously, benoxaprofen inhibits the lipoxygenase that converts arachidonic acid to leukotrienes.. Evolutionary biologist William Garstang applied "recapitulation" to the ability of evolution to retrace its steps, to escape from a dead-end, evolutionary embryologist Sir Gavin de Beer likening the process to the rewinding of a biological clock.

Evolutionary biologist Michael Whiting has shown that on a number of occasions in the past 300 million years, stick insects lost and re-evolved their wings. Five adult siblings that walk only on all fours were discovered in Turkey. It is possible that they never evolved into bipeds, or once bipeds regressed to quadrupeds.

## Prostaglandin homologies

The term "homologue organ" was coined by evolutionary zoologist Geoffrey St. Hilaire, who wrote, "It is not generally acknowledged that vertebrates are built up on one uniform plan - e.g., the forelimb may be modified for running, climbing, swimming or flying, yet the arrangement of bones remains the same ... Homologous structures share the same evolutionary origin. Mammalian limbs are all constructed on the same basic plan, as are the mouthparts of insects." Darwin held that homology is a product of natural selection. By making small cumulative changes, natural selection can transform a leg into a wing or a flipper. Homologies are of form and function, and reinforce the theories of evolution and natural selection.

For female Rheobatrachus frogs to brood their eggs in their stomachs, some factor must turn off the secretion of hydrochloric acid and suppress the passage of eggs into the intestine. Tyler, O'Brian and coworkers have shown that prostaglandin E2 secreted by brooded tadpoles inhibits acid secretion in the stomach of the female until the tadpoles have completed development and emerge as juvenile frogs through the females mouth.

At hatching, when tadpoles enter the water, their external gills rapidly regress. Evolutionary biologist Karen Warkentin has shown that the loss of

these gills is a response to increased oxygen availability. Based on the homology of the mammalian
ductus arteriosus with a portion of the amphibian
brachial arches, and the regulation of blood flow
in the mammalian ductus by prostaglandins of the
E family, Warkentin and Evolutionary herpetologist
Richard Wassersug hypothesized that anuran gill loss
is also regulated by prostaglandins. They applied
prostaglandin E2 and a synthetic analog of prostaglandin E1 (misoprostil), to embryos and young
hatchlings of the red-eyed tree frog. Both agents
accelerated external gill regression. Warkentin and
Wassersug suggest that the source of the prostaglandins may be the oral mucus of the tadpoles.

In 1991 H. Toh brought to light the evolution
of the enzymes in the cyclooxygenase and lipoxygenase pathways of arachidonic acid metabolism. In 1995, evolutionary biologist M E Baker
proposed that the genealogy of regulation of
human sex and adrenal function, prostaglandin
action, snapdragon and petunia flower colors,
antibiotics and nitrogen fixation derive from two
ancestral dehydrogenases, one of which regu-
lates prostaglandin E2, the other a common
ancestor with proteins in a soil bacterium that
forms nodules on the roots of legumes, and takes
up nitrogen from the atmosphere.

The highest concentrations of prostaglandins
in nature are found in coral. D R Koljak at the
Tallinn Technical University in Estonia has isolated
an ancestor of cyclooxygenase that is responsible
for prostaglandin synthesis in coral. They propose

that vertebrate cyclooxygenases are evolutionary derivatives of the invertebrate precursor. At the University of Aberdeen, J Zou has shown that cyclooxygenases have a long evolutionary history, probably dating back to the evolution of fish over 55 million years ago.

## Prostaglandins in Metamorphosis (a complete change of form, structure, or substance).

Wassersug addresses the timing of metamorphosis of tadpoles into adults. He suggests that as free-living tadpoles excrete prostaglandins of the E families in their oral mucus, these compounds are swallowed with food particles. When food is abundant larvae swallow a large amount of hormone that retards differentiation of the adult, acid secreting peristaltic stomach. When less food is available, less prostaglandin transits the alimentary tract and the gut proceeds to differentiate.

Jennifer Sheridan and David Bickford report that forty-five percent of species studied reach smaller adult sizes than they did recently. They invoke climate change induced warmer temperatures and changing habitats as possible culprits. The changes were most pronounced in cold-blooded animals and fish, but warm-blooded animals such as birds, mammals, and polar bears were not spared. It may well be that natural selection and variation are responsible for these anomalies, and natural selection for global warming.

## Sex Behavior

Neurobiologist Christopher Wright notes, "The body of evidence suggests that ....prostaglandin receptors.... are both necessary and sufficient for PGE2-induced masculinization of sex behaviors.... McCarthy illuminates the role of prostaglandins in the masculinization of the preoptic area of the brain and control of male sexual behavior.

## Prostaglandins in apoptosis

Apoptosis, or programmed cell death, is essential to growth and development, to the pruning and culling of neurons in the central nervous system. Apoptosis is, paradoxically, involved in infectious disorders, autoimmune disorders and cancer. Prostaglandins and their enzymes are responsible not only for apoptosis, but also for its paradoxical qualities.

## Prostaglandins in death and extinction

In 1818 George Cuvier showed that the fossil record contains extinct species. Most of the species that have ever lived are extinct. Darwin wrote that evolutionary change by natural selection requires the elimination of inferior varieties. Extinction may be due to a series of events or one catastrophe. Within each species life is possible within a range of concentrations of prostaglandin s, impossible out-side of these values. Prostaglandins are responsi-ble for so many potentially fatal diseases that they may induce death itself. As Darwin commented, "That, which creates life, destroys life."

# LITHIUM SALTS

Microorganisms, changes in availability of specific nutrients, alterations in the composition of the environment, or time may have induced such increases in prostaglandins in various species as to be incompatible with life. A naturally occurring substance that limits increases in a prostaglandin could disappear from the food or water supply or be lost due to migration. Lithium is one such element.

There is a large gap in the fossil record of human evolution during the Pliocene period. The seas were rich with shellfish, mollusks, sea urchins and bony fish. On land, life diversified and specialized into saber-toothed cats, huge bears and hyena-like dogs. Primates, includ-ing monkeys, apes and gibbons became more specialized. One result of the primate speciali-zation was Australopithecus, the ancestor of modern man.

During these 3.7 million years Australopithecus may have thrived at the seashore, where sea-weed, shellfish and animal marine life was plentiful. Lithium is present in relatively large amounts in shellfish and other seafood. Lithium would have given Australopithecus a selective advantage by stimulating his immune system to defeat various viruses and bacteria, thus able to survive epidemics that may have wiped out species not ingesting it. Lithium would also have had pacifying and socializing effects on early man. If so, reduction in lithium in available food and water would have paradoxically released both our creativity and penchant for violence and war, while exposing us to many of our disorders of defective immunity. Research in the early 1970's showed that lithium inhibits prostaglandin E1. Recent work adds that lithium acts on phospholipases and increases the turnover of arachidonic acid in the brain. It would be surprising if prostaglandins, and the enzymes that synthesize them, were not involved in the catastrophic extinction of species.

# PROSTAGLANDINS REGULAT

| | Reproduction |
|---|---|
| Replication | |
| Inheritance | Variation |
| Mutation | Interaction with the environment |
| Signaling | Migration |
| Differentiation | Shape |
| Form | Folding |
| Homology | Neoteny |
| Heterochrony | Metamorphosis |
| Recapitulation | Aging |
| Formation of cytoskeleton | Extinction |
| The synthesis and expression of genes | Death |

# INFINITESIMAL, EPHEMERAL AND POWERFUL

After I developed the hypothesis that prostaglandins are molecules of natural selection and evolution, I realized that every prostaglandin molecule must possess these properties. Evolution crystallized from an intangible, metaphysical force into molecules largely composed of carbon, hydrogen and oxygen. Hidden from all but the most sensitive assays, these molecules determine whether or not an organism reproduces, raises its young, and contributes to the evolutionary pathway of its species.

The theory of natural selection holds that minute variations determine whether or not individuals survive and reproduce. Infinitesimal prostaglandins determine reproductive failure or success, and the ability of the offspring to adapt to the environment and survive to reproduce.

## PROSTAGLANDINS: SIGNALING MASS DEATHS

"Discovery consists of seeing what everybody has seen and thinking what nobody has thought"

Albert St. Gyorgy

In Arkansas, thousands of red-winged blackbirds dropped dead out of the sky. 3000 drum fish died in the Arkansas River, 5000 birds in Louisiana and Kentucky. Pundits guessed that the birds died of blunt trauma to their organs, resulting from a midair collision instigated by the sound of fireworks, a die off of turtle doves to overfeeding and indigestion, of fish to a lack of oxygen, of dead cows to stale sweet potatoes. Devil crabs, sardines, croaker, catfish, bream, carp, roach fish, starlings, cowbirds, and jackdaw crows perished in their hundreds, thousands or millions. More than two hundred cows dropped dead in a field in Wisconsin, seven thousand water buffalo in Vietnam.

The blackbirds showed no signs of trauma or infection, but did have striking evidence of bleeding and clotting, a giveaway that thromboxane synthase, the enzyme that produces thromboxane B2, was induced by environmental stress, the enzyme a variant that Darwin and Wallace had in mind. In humans, this enzyme separates those susceptible to heart attacks and strokes from those that are not. Induction of cyclooxygenase, or inhibition of the prostaglandin degrading enzyme, would have been responsible for defective immunity in white snouted bats, and induction of lipoxygenase in acute respiratory deaths. Every death due to natural causes in nature probably occurs when prostaglandins reach levels that are incompatible with organ function.

Myers and Ramwell note that in arachidonic acid-induced sudden death, agents that either inhibit thromboxane generation or block thromboxane receptor activation prevent the occurrence of thrombotic death, antidepressants among them. Arachidonic acid is released by activation of phosphorylases.

The environmental factors relevant to heart attacks include exposure to heat, cold, emotional stress, battle, particulate matter, secondhand smoke, and noise pollution. Prostaglandins and thromboxanes mediate all, leukotrienes nasal blockage and respiratory failure. Whatever the environmental factor, the final common pathway in the mass deaths would have been activation of arachidonic acid cascade enzymes. Emil Zuckerkandl and Linus Pauling advanced the idea that enzymes they referred to as "semantides", are biological clocks, changing slowly over time. Of the altered physical conditions responsible for the mass deaths, rapid movement of the Magnetic North Pole towards Russia has the edge.

## SEISMIC ACTIVITY (EARTHQUAKES).

The mass deaths of fish in Japan and China that preceded earthquakes raise a tantalizing question concerning the role of natural selection in seismic activity. Unusual activity in dogs, cats, goats, horses, birds, and cows has occurred prior to an earthquake, all attributed to a hypersensitivity to variations in the earth's magnetic field that triggered the seismic events. Many astrophysicists believe that pre-biotic natural selection was at work in the formation of the universe.

## PROSTAGLANDINS IN EVOLUTIONARY MEDICINE

Depression predisposes, among others, to infectious, neurodegenerative, autoimmune, and cardiovascular disorders. In "A Medical Revolution for the Ages" (2011, Amazon) I list medical disorders known to respond to antidepressants, thus incriminating excessive prostaglandin production in all. As prostaglandins are the agents of natural selection and evolution, it follows that virtually all of our diseases and disorders are evolutionary, and may involve heterochrony, neoteny and homology.

## THE PARADOXICAL ADVANTAGES AND DISADVANTAGES OF NATURAL SELECTION: THE CASE HISTORY OF CHARLES DARWIN

The biology of natural selection was an enduring mystery, as was the nature of Charles Darwin's chronic illness. Of the theories advanced to explain the latter, oedipal conflicts and Chagas' disease are preeminent. Hypomania, however, propelled Darwin to the pinnacle of scientific achievement and good health, the depression that followed condemning him to intellectual stagnation, lethargy, impaired memory and concentration, and incapacitating gastrointestinal disorders. Examples of natural selection in humans are much sought after when, ironically, one need look no further than Darwin himself.

Arthur Koestler was generations ahead of his time when he commented, ''I've tried to account for man's creativity and his genocidal madness, the two sides of the coin minted by the evolutionary process". Koestler did not realize that bipolar disorder is the intervening variable. The role of mood disorders in creativity, genius, defective immunity and autoimmunity illuminates the enduring mystery of the nature of Darwin's chronic illness. Several writers have linked bipolar disorder to high achievement. In ''the Psychology of Men of Genius'' (1931) Ernst Kretschmer identified bipolar disorder in men of genius and their families. He observed that there is a resemblance between mania and periods of creative productivity, while depression is similar to periods of sterility. He concluded that while severe states of bipolar disorder are counter-productive, talent alone could not achieve genius without assistance from the milder states of the disorder. In ''Some Reflections on Genius'' (1960) Lord Russell Brain retreated from Kretschmer's position, stating only the bipolar disorder and genius are ''closely related''. Nancy Andreassen, who studied fifteen writers, confirmed Kretschmer's conclusion that severe states of the illness interfered with work or decreased its quality. She found that milder manic and depressive states enhance specific aspects of creativity. She wrote ''a variety of artists, writers, statesmen, philosophers, and scientists have suffered from disorders of mood''. In ''Moodswing'', Ronald Fieve commented that mild mania enhances creativity, and that many eminent people in various fields have been bipolar. Many other writers have explored the influence of bipolar disorder on the life and work of creative geniuses. In ''A Brotherhood of Tyrants: Manic Depression and Absolute Power''

D Jablow Hershman and I depict the paradoxical role of the disorder in the lives and behavior of such tyrants as Napoleon Bonaparte, Adolf Hitler and Joseph Stalin.

That bipolar disorder can produce both geniuses and tyrants is one of its paradoxes. The disorder can be a disabling illness that prevents the accomplishment of anything, and it can make life unbearable. The romantic poet Heine asked: ''what is the real reason for the curse that hangs over all men of great genius?''. While it does not hang over all of them, there is some truth in the romantic idea that genius is paid for by suffering. The suffering, however, is caused by bipolar disorder, not by the possession of talent per se. The illness has positive as well as negative effects on the lives and works of creative people. It can enhance the talents of one person, and destroy the productive capacity of one equally talented. The outcome depends largely on the severity of the illness and the course it follows. But, for a few, during some part of their lives, the disorder does confer assets that can lead to incomparable achievements. Those in scholarly professions receive from hypomania a general heightening of intellectual processes, including the ability to remember whatever they need, an abundance of insights and original ideas, seemingly effortless comprehension, and the capacity to construct and work with complex structures of thought. Hypomania also bestows phenomenal energy and an insistent urge to do something.

In depression, intellectual processes may become impaired and slowed down, memory, the capacity to solve problems and to generate ideas, comprehension, the ability to think, and even the ability to form complete sentences eventually become minimal. Until he reaches an extreme state, the depressive is still able to function and can perform what is routine and mechanical, but the quality of creative work done in this state is relatively poor. The depressive feels lethargic, tires quickly, and needs more rest and sleep. He loses the will to work and becomes reluctant to exert any effort whatsoever. Various authors have proposed that bipolar disorder has an evolutionary basis.

## NATURAL SELECTION AND CHARLES DARWIN: HYPOMANIA, DEPRESSION AND CHRONIC ILLNESS

As a young man Darwin had remarkable stamina. He was industrious, observant, curious and intellectually honest. His hypomania included periods of intense mental activity, intellectual energy, and racing thoughts. He weathered cramped quarters and an ill-tempered captain on the "Beagle", explored the Andes, and rode horses with fiery gauchos on the pampas of Argentina. A biographer writes, "as Darwin rode on to Buenos Aires, even the gauchos were astonished at his energy . . . he never seems to be tired, never loses his curiosity or sense of wonder''. Darwin was immune to illness, save for the seasickness that plagued him aboard the ''Beagle''.

1836–1838 were the most active years in Darwin's life, a ferment of observing, creative reflection, writing and editing. A few months after the return of the ''Beagle'' Darwin developed chronic depression and a host of associated medical disorders that ultimately reduced him to invalidism. His symptoms of depression included anxiety, guilt, weakness, lethargy, insomnia, impaired memory and concentration, loss of confidence, and fear of dying. The medical disorders associated with depression included trembling, tingling, nausea, bouts of vomiting, chronic gastric and intestinal pain, fatigue, flatulence, palpitations, dizziness, headaches, boils, eczema and feeling cold.

Darwin was aware of his limitations. In his autobiography, he wrote: ''I have no great quickness of apprehension, which is so remarkable in some clever men, for instance Huxley. I'm therefore a poor critic: a paper or book when first read, generally excites my apparition, and is only after considerable reflection that I perceive the weak points. My memory is extensive, yet hazy: it suffices to make me cautious by vaguely telling me that I have observed or read something opposed to the conclusion which I am drawing, or on the other hand in favor of it; and after a time I can generally recollect where to search for my authority. So poor in one sense is my memory, that I have never been able to remember for more than a few days a single date or a line of poetry. My power to follow a long and purely abstract train of thought is very limited; I should, moreover, never have succeeded with metaphysics or mathematics."

Darwin consulted the hydropathist James Manby Gully, whose theory was to draw blood away from inflamed organs. Sweating would be followed by cold baths to "rouse" the system, while bland foods would not offend the stomach. Other interventions included wet sheet packing and wrapping, steam baths, and drinking spring water.

## COSMIC ANCESTRY

Lord Kelvin (1824-1907) established absolute zero, helped to devise the first undersea communications cable, formulated the first two laws of thermodynamics, transformed physics into a field of science, and estimated the age of the earth as up to 400 million years. Hermann von Helmholtz (1821–1894) was a physician that made major contributions, among others, to geometry, physics, the field theory of physics, and invented the ophthalmoscope. Both proposed natural panspermia as a theory that the universe is full of spores that germinate when they find a favorable environment. Svante Arrhenius, winner of the Nobel Prize for chemistry in 1903, theorized that bacterial spores, propelled through space, were the seeds of life on Earth.

# CARBON ABOARD METEORITES

Fernando and Rowe propose that chemical evolution can take place by natural selection in the absence of nucleic acids. Carbonaceous chondrite meteorites were thought to contain both carboxylic and amino acids, until a team headed by Sandra Pizzarello, a scientist at Arizona State University, found carboxylic acids in a meteorite with a scarcity of amino acids, this meteorite evolving before others with amino acids. Pizzarello notes "a different outcome" of organic chemical evolution in space is likely to have happened during the formation and development of the solar system, "but one that still might have contributed molecular precursors of biomolecules to the origins of life."Meteorites may have embedded carbon in the earth, carbon acquiring oxygen and hydrogen by natural selection, the end products such carboxylic acids as arachidonic acid.

Richard Hoover, Ph.D, an accomplished NASA astrobiologist, claims to have discovered evidence of microfossils similar to a bacterium in the interior surfaces of carbonaceous meteorites, and believes they are indigenous to meteors. He concludes that these bacteria are not contaminants, but remnants of living organisms which lived in such parent bodies of meteors as comets, moons, and other astral bodies.

# NITROGEN AND PHOSPHORUS

Nitrogen, which occupies 80% of the Earth's atmosphere, is an essential ingredient in amino acids, the elements of proteins, nucleic acids (DNA and RNA) and enzymes. Astrochemists have shown that the earth's early atmosphere was seeded by nitrogen-containing comets.

Phosphorylation refers to the combination of phosphate with such carboxylic acids as arachidonic acid, thereby creating phospholipids- the fundamental components of cell membranes, which subserve metabolism, signaling, and transport of molecules into the cell. Once enclosed, cells could replicate and form primitive microorganisms lacking in nitrogen. Arachidonic acid transformed nitrogen into amino acids, proteins and enzymes, and ultimately nucleic acids, chromosomes and genes.

# LITHIUM ABOARD METEORITES

The Spite lithium plateau is a baseline in the abundance of lithium found in old stars. It was named after the astronomers François and Monique Spite, who published their discovery in 1982. Lithium was produced during collisions of cosmic rays, or the evolution of stars. It was not a component of the earth's original mantle, but hitched a ride aboard meteorites.

Lithium does not occur "free" (purified or uncombined) in nature, but is found in igneous rocks, and in many mineral springs and seawater. Traces of lithium are found in numerous plants, plankton, and invertebrates. Only near hydrothermal vents under the oceans does seawater contain elevated levels of lithium. .

Lithium controls crystal shape in synthetic crystals of the calcium carbonate mineral Aragonite. This property is ... reminiscent of the structure observed in the shells of many mollusks.... the lithium ion has .....Profound biological effects of varying intensity on the gamut of life forms, have confirmed its influence on enzyme activity, metabolism, respiration, and active transport.

In the late nineteenth century, Herbst showed that when lithium is added to sea water, it profoundly modifies sea urchin development. Today, lithium is known to cause structural changes throughout the animal kingdom, but most conspicuously in primitive embryo's. Such changes have been observed in tunicates (sea squirts), cyclostomes (lampreys, hagfish), teleosts, (most living fish species) and cephalopods (octopus, squid, cuttlefish).

*Excerpted and adapted from*
*"Some Facts about Lithium."*

In animal studies, lithium has a role in the expansion of the stem cell pool to blood cells. Oliveira has shown that lithium influences enzymes and signaling in the brains of zebrafish. As an immunostimulant, antiviral and antibacterial agent, lithium might have played an essential role in human embryogenesis; and an indirect role in human evolution in the embryogenesis of seafood, thus proving a rich source of nutrition. Lithium was not part of the earth's original mantle, but arrived here in meteorites, the fragments of explosions of population 11 stars, thus adding to the plausibility of cosmic seeding of life on earth, minable gold recently now accorded immigrant status.

## Oceans from Comets

The European Space Agency's Herschel infrared space observatory has found water in a comet with almost exactly the same composition as Earth's oceans. The discovery revives the idea that giant icebergs floating through space may have formed our seas.

# ELECTRIFICATION

Modern high tech science has shown that prostaglandins are electrified. In 1894, using relatively primitive appaaratus, Lord Kelvin demonstrated the electrification of carboxylic acid, now known to be the chemical structure of prostaglandins. Electromagnetic fields, generated by earths liquid iron core, bombarded the earth with electrically charged particles would have electrified carboxylic acids, with lightning a distant possibility.

Electromagnetic fields stimulate and inhibit prostaglandin E2, while electrophilic prostaglandins may modify proteins. Dr B Song has shown that electrical cues regulate the orientation and frequency of cell division and the rate of wound healing, as do prostaglandins, and prostaglandins sensitize nerve cells to electrical stimuli. This would explain why stubbing a toe can light up the brain in milliseconds, and a magnetic pole shift synchronously fire up thromboxane synthase in the platelets of red- winged blackbirds.

# MASS DEATHS

Fast forward by billions of years to the fall of 1978, when I observed remission of herpes virus infections in two patients taking lithium for mood regulation. That chance event evolved one small piece at a time, into the hypothesis that lithium has antiprostaglandin, immunostimulating and antiviral and antibacterial properties, antidepressants antiprostaglandin, immunostimulating, antiviral, antibacterial, antiparasite, and fungicidal properties. Then backwards in time, to studies published in the mid-seventies, showing that prostaglandins, carbon- based carboxylic acids, regulate deoxyribonucleic acid (D.N.A) and ribonucleic acid (RNA), nitrogen based nucleic acids, thus regulating the synthesis, inhibition, and expression of genes.

This brings us closer to the presence of lithium in meteorites, and mass deaths of humans, animals, birds, fish, crabs, sardines, dolphins, cows, water buffalo, crickets, seals and bees. Nature's carboxylic acid immune defenses respond to all environmental challenges, the outcomes determined by the ability of these acids to withstand the stress. Should the stressor induce carboxylic acids beyond a critical threshold, immune function will weaken, the host challenged by acute or chronic illness. Lithium inhibits the phosphorylase enzymes that generate arachidonic acid, and would give the edge to primitive nitrogen containing bacteria with phosphorylase enzymes, more evolved by the presence of nitrogen. Human mass deaths are either acute and preventable with lithium, or chronic and preventable with antidepressants.

I was puzzled by the relationship of the Jasper tornado to fungal infections of the skin, until realizing that the fungus was dormant, the force provoking the tornado also depressing immune function in the individuals saddled with skin lesions. This suggests that the fungus incriminated in the white nosed bat syndrome was dormant, until natural selection stimulated cyclooxygenase in vulnerable bats into depressing immune function. The hypothesis that arachidonic acid, and the enzymes of its cascade are the driving force of natural selection and evolution is on solid ground, the interface between prostaglandins and nucleic acids of paramount importance in biomedical research. .

Mass extinctions, including those of dinosaurs and mastodons, are usually attributed to meteorites, rising sea levels, volcanoes, glaciations and comet showers. The sixth mass extinction, now in progress, is alleged to be the result of destruction of ecosystems, plundering of natural resources, overpopulation, agriculture and global warming.

The theory of uniformitarianism holds that all geologic phenomena may be explained as the result of existing forces having operated uniformly from the origin of the earth to the present time. Electrification of carboxylic acids suggests that electromagnetic fields orchestrated the origin of species and their extinction, or in Darwin's' words, "That which creates life destroys it." The enzymes of the arachidonic acid cascade and the enzymes of the replication of nucleic acids have always differentiated between reproduction and sterility, and between life and extinction.

---

The medical content of this book is published under the provisions of article 19 of the universal declaration of human rights of 1946, which were intended to prevent atrocities such as occurred during the Second World War. It is my ethical duty, human, right, and professional responsibility to disseminate the information as widely as possible, to patients, physicians, and citizens, and especially the impoverished, ailing, disabled or disadvantaged in any other manner. Human rights prohibit evaluation or distortion of the information, and delay in disseminating it. Readers are encouraged to access databases for more information. It is an advantage to have a sense of humor for those brave enough to access Pubmed central for "Nucleic Acids Research Vols. 1 to 39; 1974 to 2011." This book is for informational purposes only. All treatment decisions to be made with a physician.

www.ingramcontent.com/pod-product-compliance
Lightning Source LLC
Chambersburg PA
CBHW060229290526
45789CB00003B/1481